RAND JUSTICE, INFRASTRUCTURE, AND ENVIRONMENT

Rethinking Insurance and Liability in the Transformative Age of Autonomous Vehicles

James M. Anderson, Nidhi Kalra, Karlyn D. Stanley, Jamie Morikawa

Preface

The potential benefits of autonomous vehicle (AV) technology are described in media reports almost daily. Many new model vehicles on U.S. roads include aspects of AV technology, such as lane-changing alerts, forward collision warning, and blind-spot warnings. As fully autonomous vehicle technology begins to be deployed in more vehicles in the United States, the auto insurance industry also will enter a time of transition. The liability regime that currently underpins auto insurance, which is based on individual driver liability, may soon change to reflect a greater emphasis on the liability of the manufacturer of an AV. In the event of a car crash, if a person in an AV was not actually driving, the question will be: Who should be held responsible?

On July 14, 2017, the RAND Institute for Civil Justice held a workshop titled "Rethinking Insurance and Liability in the Transformative Age of Autonomous Vehicles" in San Francisco, which brought together stakeholders including insurers, reinsurers, plaintiff's attorneys, defense attorneys, automotive industry representatives, regulators, and consumer representatives. The discussion focused on the implications of AV technology for insurance and liability regimes as we enter a time of transition. The workshop addressed key uncertainties and challenges for the existing liability and insurance landscapes and explored how these systems might address any potential disruptions. The workshop also addressed what the effects on consumers might be. The event was sponsored by the law firm of O'Melveny and Myers and the American Insurance Association. This conference proceeding summarizes key issues and topics from the workshop. The document is not intended to be a transcript; rather, it organizes the major themes of discussion by topic, pointing out areas of agreement and disagreement among participants. To encourage open dialogue, the authors of this document elected not to attribute specific remarks to individual participants or panelists, with the exception of the RAND speakers.

These proceedings should be of interest to stakeholders in the insurance and reinsurance industries, plaintiff's attorneys, defense attorneys, corporate executives in the automobile and artificial intelligence industries, policymakers and regulators who are grappling with new and emerging technologies, and other stakeholders with an interest in the implications of AV technology for insurance and liability regimes. In addition to these proceedings, stakeholders may be interested in related RAND research concerning AVs, including *Autonomous Vehicle Technology: A Guide for Policymakers* (Anderson et al., 2016), *The Enemy of the Good: Estimating the Cost of Waiting for Nearly Perfect Automated Vehicles* (Kalra and Groves, 2017), and *Autonomous Vehicles and Federal Safety Standards: An Exemption to the Rule?* (Fraade-Blanar and Kalra, 2017).

The RAND Institute for Civil Justice

The RAND Institute for Civil Justice (ICJ) is dedicated to improving the civil justice system by supplying policymakers and the public with rigorous and nonpartisan research. Its studies identify trends in litigation and inform policy choices about liability, compensation, regulation, risk management, and insurance. The institute builds on a long tradition of RAND Corporation research characterized by an interdisciplinary, empirical approach to public policy issues and rigorous standards of quality, objectivity, and independence.

ICJ research is supported by pooled grants from a range of sources, including corporations, trade and professional associations, individuals, government agencies, and private foundations. All its reports are subject to peer review and disseminated widely to policymakers, practitioners in law and business, other researchers, and the public.

ICJ is part of RAND Justice Policy within RAND Justice, Infrastructure, and Environment, a division of the RAND Corporation that conducts research and analysis in civil and criminal justice, infrastructure development and financing, environmental policy, transportation planning and technology, immigration and border protection, public and occupational safety, energy policy, science and innovation policy, space, telecommunications, and trends and implications of artificial intelligence and other computational technologies.

Questions or comments about this report should be sent to the director of the RAND Institute for Civil Justice, James Anderson (janderson@rand.org). For more information about the RAND Institute of Civil Justice, see www.rand.org/icj.

Contents

Summary

Autonomous vehicle (AV) technology offers the possibility of dramatically reducing fatal car crashes and crippling injuries. Statistics published by the National Highway Traffic Safety Administration (NHTSA) show that, "[i]n 2016, motor vehicle-related crashes on U.S. highways claimed 37,461 lives" (NHTSA, undated). NHTSA's research concluded that "94 percent of serious crashes are due to human error" (NHTSA, undated). The same NHTSA document showed "motor vehicle crashes in 2010 . . . cost $594 billion due to loss of life and decreased quality of life due to injuries" (NHTSA, undated). Automobile insurance has been developed to address the economic costs of car crashes and the resulting fatalities and injuries. As fully autonomous vehicle technology begins to be deployed in more vehicles in the United States, the automobile insurance industry also will enter a time of transition. The liability regime that currently underpins auto insurance, which is based on individual driver liability, may soon change to reflect a greater emphasis on the liability of the manufacturer of an AV. In the event of a car crash, if a person in an AV was not actually driving, the question will be: Who should be held responsible?

When will fully autonomous vehicles be on the road? There is uncertainty about the actual deployment of full AV technology, although by 2020, there may be pilot projects involving the public. NHTSA (undated) shows that 2025 is likely to be the advent of fully automated safety features and highway autopilot. Currently, many new car models incorporate advanced driver-assistance systems that are part of AV technology, such as forward collision warning, automatic emergency braking, pedestrian automatic emergency braking, adaptive lighting, adaptive cruise control, lane departure warnings, rearview video systems, and rear cross-traffic alert. Although these new systems may improve the safety of new car models, it is likely that there will be a mixed fleet of AV and non-AV vehicles for decades. Therefore, it is unclear how and when AV technology may affect the automobile insurance industry.

It is in this context that RAND's Institute for Civil Justice held a workshop on July 14, 2017, titled "Rethinking Insurance and Liability in the Transformative Age of Autonomous Vehicles," to stimulate a pathbreaking conversation about the challenges and opportunities facing the automobile insurance industry when AV technology becomes widely deployed. In addition, RAND researchers hoped to identify important areas for study in this time of transition. The workshop brought together stakeholders, including insurers, reinsurers, plaintiff's attorneys, defense attorneys, automotive industry representatives, regulators, and consumer representatives, many of whom are already dealing with aspects of AV technology. The workshop required participants to address key uncertainties for the existing liability and insurance landscapes and explore how these systems might respond to any potential disruptions.

Several major themes emerged during the workshop. First was the consensus recognition that, for a variety of reasons, AV technology may not be transformative for liability and insurance regimes in the near term. For example, workshop participants discussed the presence of 260 million non-AVs on the road today and the fact that they will not go away overnight. The Insurance Institute for Highway Safety predicts a three-decade adoption period between the introduction of a safety feature (e.g., antilock brakes) and its adoption. Currently, AVs are exclusively corporate, and there is no individual ownership of AVs. There was general agreement among the participants that, until AVs are widespread among the public, there will not be changes in insurance industry standards. Such a shift could mean a change from operator to manufacturer liability—but participants in the workshop agreed that more information was needed to make such a prediction.

AV data access and sharing was another theme. A workshop participant stated that the number-one public policy issue for the insurance industry is the relevance and importance of data access. The group agreed that all stakeholders are thinking about data—and everyone is guarding their data. It is unclear what role, if any, the government will play in the AV data-sharing process. Privacy will be a key issue, and whether and how data can be successfully de-identified will be critical.

The security of AVs from "hacking" by a malicious actor emerged as an important theme. There are dangers posed by massive hacks of AVs, as well as private hacks, which are random and more isolated. Possibly, insurers can identify opportunities for new insurance products. Currently, auto insurance policies may not cover either massive or individual hacking of AVs.

A final theme was the need for the insurance industry to be proactive, not reactive. The group discussed the possibility of developing new insurance coverage models and the need for a working group to further these efforts. A working group might identify and address the types of international cooperation that would be possible. Generally, the group thought it was important to find a way to modernize regulations to better deal with industry-altering technologies, such as AVs.

Morning Session: Key Uncertainties That Will Shape Future Liability for Autonomous Vehicles

The morning session began with an introduction by James Anderson, director of the RAND Institute for Civil Justice, about the goals of the workshop. The introduction was followed by a presentation by Nidhi Kalra, director of RAND's San Francisco Bay Area office, that addressed the state of AV technology, as well as its potential advantages and disadvantages. The presentation explained that, by 2020, there may be pilot projects involving the public. However, the introduction of AVs will lead to a time of uncertainty, particularly concerning policies that currently regulate the automotive and insurance industries. This presentation set the stage for participants in the workshop to engage in a roundtable discussion with panelists in the next session.

The moderated panel discussion that followed the introduction included a representative of a ride-sharing company, a plaintiff's lawyer, a former U.S. Department of Transportation regulator, and an insurance industry expert. The panelists were asked a series of provocative questions to identify some of the uncertainties that may shape the future of liability and insurance for AVs. These questions included:

- whether AVs were a "game changer," particularly concerning liability and insurance regimes
- whether there was an anti-machine bias that would require AVs to be more-perfect drivers than humans
- what amount of testing is needed to determine the safety of AVs and the expected timeline of the regulatory process
- what are the biggest issues in the insurance industry related to AVs
- whether there is an emerging conflict between AV proprietary data and data that the insurance industry needs.

In response to the question about whether AVs were a game changer, the panelists concluded—for different reasons—that AV technology may not be transformative for liability and insurance regimes in the near term.

The moderator addressed the second question by suggesting that, as a society, humans accept that they are not perfect—that they make mistakes. She noted that, as a society, we seem to have a problem accepting that machines also make mistakes. She asked the panelists whether they thought that the manufacturers and creators of AVs are going to be at an unfair disadvantage in the court room, stemming from an anti-machine bias. The panelists responded that there are likely to be very high expectations for AV technology on the part of consumers for a variety of reasons, which they discussed. The panel concurred that, if the societal expectation is that AV technology is going to save lives, society is going to hold the manufacturer to that standard. If purveyors of AVs make statements about safety, then people will expect AVs to be safe.

The panel discussed the third question—what is "safe enough" for AV deployment—with varied perspectives. Some panelists agreed that more people will adopt AV technology than we expect—and more quickly. Another panelist countered that the difference with AVs is that people will expect complete safety and will not think they are assuming any risk.

The discussion turned to the fourth question concerning the regulatory process for AVs and the estimated timeline for regulatory changes. The panel identified factors that could impact the regulatory timeline, including interaction between Congress and regulatory agencies, the proliferation of current proposed legislation about AVs on Capitol Hill, and the development of standards by NHTSA.

Finally, the panelists were asked to identify the biggest issues in the insurance industry related to AVs. The issues identified included the inability of the insurance industry to obtain data to measure aspects of AV technology, the need for regulatory changes, and the potential need to insure against AV hacking by malicious actors.

Afternoon Breakout Sessions: Liability and Insurance Rethink

In the afternoon, the participants were divided into two smaller working groups designed to facilitate deeper conversation about liability and insurance for AVs. Both groups were asked to identify important research questions in these areas. A consolidated summary of the issues discussed by both groups is provided in this section.

There was a robust discussion about how insurance companies should determine the cost of AV insurance plans and decide how to pay out following AV accidents. The group discussed the tensions between sharing data for safety and insurance reasons and protecting consumer privacy. One of the most important privacy issues is de-identifying location and geographic data. Data sharing between manufacturers and insurance companies is necessary for insurance pricing; data sharing between manufacturers and courts is necessary for litigation; and data sharing between manufacturers and law enforcement is necessary for security and crime prevention and prosecution.

The group agreed that it was important to understand the state of the art of artificial intelligence (AI). A key concern was the lack of traceability of AI. The group discussed the need for manufacturer documentation of AI used in an AV, as well as the potential for strict liability because of low traceability.

The conventional wisdom is that driving jobs will disappear with the full deployment of AVs, but the group questioned whether this was accurate. Participants debated whether AVs might increase opportunity for more-experienced drivers, such as taxi drivers.

The group discussed the probability that the insurance industry will not be able to insure AVs in California because of Proposition 103, which requires that insurers focus on three factors when determining the cost of coverage: (1) driving record, (2) number of miles the driver has driven, and (3) length of time the driver has been driving. Clearly, these criteria are inapplicable in an autonomous vehicle world.

Infrastructure was identified as playing an essential role in AV deployment. The group identified grading roads, optimizing road conditions, improving signage and striping, and creating a standard classification system as important infrastructure issues.

Finally, the group debated the government's role in managing the competing types of technologies on the road. A key issue the group identified concerned whether ridesharing would increase the number of cars on the road.

Key Issues, Research Questions, and Next Steps

In the afternoon's final session, each of the working groups reported their key issues based on the group discussions and provided suggestions for further research.

The groups highlighted the following issues as the most critical:

- Insurers and manufacturers need to identify the state of the art of AV technologies going forward.
- AV manufacturers, developers, and insurers need to fully understand AI and explain to consumers how AI makes decisions.
- Insurers need AV data to create pricing policies for automobile insurance and to assess underwriting risk.
- The security of AVs, particularly in terms of hacking by a malicious actor, is a very important issue for the insurance industry.
- Original equipment manufacturers (OEMs) have powerful brands, and they are going to hesitate to admit fault in the event of an AV crash.
- The advent of new technologies might allow us to rethink how our traditional approach to regulation may not create good policy.
- Predictions of catastrophic job loss resulting from AVs may be overstated in the short term.
- It is important to "think beyond the car" in creating a safe environment for the introduction of AVs.
- It may be difficult to insure AVs in California under the current regime unless Proposition 103 is changed.

The two groups identified many issues for research. These included the following:

- Investigate and better understand consumer attitudes about trusting and adopting a new technology.
- Create a research design to evaluate consumers' use of AV technology as soon as it is available to consumers.
- Conduct research to explore other industries that have experience with AI (e.g., aviation, nuclear engineering) to identify how the AV industry might learn from them.
- Evaluate whether driving culture varies by region and geography and whether AV algorithms will need to drive differently in different regions.
- Investigate the AV data that are needed by consumers, manufacturers, regulators, and insurers.
- Conduct research to identify and quantify the impact of AVs on insurance coverage.
- Evaluate cyber safety and privacy issues posed by AVs.

The workshop concluded with a final roundtable discussion prompted by the research questions identified by both groups. Among the key ideas for next steps that emerged in this discussion were suggestions that the insurance industry needs to be proactive, not reactive. The group discussed the possibility of developing new insurance coverage models and the need for a working group to further these efforts. The group concurred that consumer acceptance and education would be essential to successful widespread deployment of AV technology. The participants agreed that they would like to find a way to continue the conversations that emerged in the day's workshop.

Acknowledgments

We would like to thank the panelists, participants, moderators, and all those who engaged in the roundtable discussions, without whom the exchange of ideas documented here would not have been possible. We would also like to thank our sponsors, O'Melveny and Myers and the American Insurance Association, for their generous support of the workshop. We would like to recognize Daniel Dunmoyer, chair of the RAND Institute for Civil Justice Board of Overseers, for suggesting the idea of the workshop and for his leadership of the Board of Overseers.

The views expressed during the event represent the perspectives of the various participants and do not necessarily reflect the views of their employers or the sponsors of the event.

Abbreviations

AI	artificial intelligence
AV	autonomous vehicle
FMVSS	Federal Motor Vehicle Safety Standards
NCSL	National Conference of State Legislatures
NHTSA	National Highway Traffic Safety Administration
OEM	original equipment manufacturers
TNC	transportation network companies

1. Introduction

The expected time frame for the deployment of fully autonomous vehicles (AVs) is uncertain for both technological and political reasons. AV technology can be conceptualized using a five-part continuum developed by the Society of Automotive Engineers (SAE) and adopted by the National Highway Traffic Safety Administration (NHTSA). The stages of autonomy are described on NHTSA's web site (NHTSA, undated):

- Level 0: No Automation. Zero automation, the driver performs all driving tasks.
- Level 1: Driver Assistance. Vehicle is controlled by the driver, but some driving assist features may be included in the design.
- Level 2: Partial Automation. Vehicle has combined automated functions, like acceleration and steering, but the driver must remain engaged with the driving task and monitor the environment at all times.
- Level 3: Conditional Automation. Driver is a necessity, but not required to monitor the environment. The driver must be ready to take control of the vehicle at all times with notice.
- Level 4: High Automation. The vehicle is capable of performing all driving functions under certain conditions. The driver may have the option to control the vehicle.
- Level 5: Full Automation. The vehicle is capable of performing all driving functions under all conditions. The driver may have the option to control the vehicle.

When will Level 5, fully autonomous vehicles, be on the road? This is uncertain, although NHTSA's website shows that 2025 is likely to be the advent of fully automated safety features and highway autopilot (NHTSA, undated). Currently, many new car models incorporate advanced driver-assistance systems that are part of AV technology, such as forward collision warning, automatic emergency braking, pedestrian automatic emergency braking, adaptive lighting, adaptive cruise control, lane departure warnings, rearview video systems, and rear cross-traffic alerts. Although these new systems may improve the safety of new car models, it is likely that there will be a mixed fleet of AV and non-AV vehicles for decades. Therefore, it is unclear how and when AV technology may impact the automobile insurance industry.

The question of when Level 5, fully autonomous vehicles, will be on the road has political—and technical—dimensions. According to the National Conference of State Legislatures (NCSL), in 2017, 33 states introduced AV legislation, and, "since 2012, at least 41 states and D.C. have considered legislation related to autonomous vehicles" (NCSL, 2018). The NCSL's webpage shows that 21 states—Alabama, Arkansas, California, Colorado, Connecticut, Florida, Georgia, Illinois, Louisiana, Michigan, New York, Nevada, North Carolina, North Dakota, Pennsylvania, South Carolina, Tennessee, Texas, Utah, Virginia, and Vermont, and Washington, D.C.—have passed legislation related to AVs (NCSL, 2018). State legislation regulates aspects of when, where, and how AVs are allowed to drive. Existing state legislation addresses a wide variety of issues such as testing of AVs, pilot projects concerning AVs, and safety standards for AVs (NCSL, 2018).

The U.S. House of Representatives passed a bill on September 6, 2017, that would establish a national framework for regulation of self-driving vehicles. The bill, H.R. 3388, prohibits states and localities from regulating the design, construction, or performance of highly automated vehicles in an effort to maintain nationwide standards. It also increases the number of exemptions from the Federal Motor Vehicle Safety Standards (FMVSS) that NHTSA may grant in a given year to allow more testing of vehicles that may not have typical car components, such as brake pedals or steering wheels.

The U.S. Senate is working on a bill as well, although it has not yet passed the full Senate (U.S. Senate, 2017). The bill, S. 1885, would largely prevent state and local governments from regulating highly autonomous vehicles but would let states and localities regulate licensing, liability, and insurance for the vehicles. The bill would also allow NHTSA to issue a comprehensive update to such regulations as the FMVSS. Senators Richard Blumenthal and Edward J. Markey are holding up the bill to push for measures they feel will improve the safety of highly autonomous vehicles, such as a provision to require AVs to allow human drivers to take control.

On July 14, 2017, the RAND Institute for Civil Justice held a workshop titled "Rethinking Insurance and Liability in the Transformative Age of Autonomous Vehicles" in San Francisco to address the issue of the impact of autonomous vehicles on the insurance industry from many perspectives. These perspectives were the subject of an expert panel discussion during the morning session. In the afternoon, the invited participants broke into working groups to identify key issues the insurance industry must address concerning the impact of AVs, as well as important areas for research.

Chapter Two of this conference proceeding summarizes the discussion during the morning session. Chapter Three provides a brief summary of remarks by a RAND expert on the issue of AV safety and policy development. Chapter Four presents the results of the afternoon working group sessions. Chapter Five outlines the key issues and research questions identified by the working groups, as well as next steps suggested by the workshop participants.

2. Morning Session: Key Uncertainties That Will Shape Future Liability for Autonomous Vehicles

Setting the Stage for Discussion

The morning opened with an introduction by James Anderson, director of the RAND Institute for Civil Justice, about the goals of the workshop. He explained that participants in the workshop would be asked to identify key issues and research questions for further inquiry into autonomous vehicle (AV) technologies, liability, and automobile insurance that developed from the day's proceedings. Following this introduction, Nidhi Kalra, director of RAND's San Francisco Bay Area Office, presented "A Tour of Our Future with Autonomous Vehicles," which addressed the state of AV technology, as well as its potential advantages and disadvantages. The presentation described an AV as one that can drive itself some of the time or all of the time, with or without human intervention. The presentation explained that there is uncertainty around the actual deployment of full AV technology, although by 2020, there may be pilot projects involving the public. Kalra explained that it is likely that there will be a mixed fleet of autonomous and nonautonomous vehicles for decades. The potential benefits of AV technology are significant: It can reduce crashes and save lives, and it can provide mobility benefits to the tens of millions of Americans who cannot drive because of disability or age. AV technology also might reduce traffic congestion and improve land use. However, the introduction of AVs will lead to a time of uncertainty, particularly concerning policies that currently regulate the automotive and insurance industries. This presentation set the stage for participants in the workshop to engage in a roundtable discussion with panelists during the next session.

Panel Discussion: Some Key Uncertainties That Will Shape Future Liability for Autonomous Vehicles

In a moderated panel discussion that included a representative of a ride-sharing company, a plaintiff's lawyer, a former regulator from the U.S. Department of Transportation, and an insurance industry expert, the panelists were asked a series of provocative questions to identify some of the uncertainties that may shape the future of liability and insurance for AVs. These questions asked

- whether AVs were a "game changer," particularly concerning liability and insurance regimes
- whether there was an anti-machine bias that would require AVs to be more-perfect drivers than humans
- what amount of testing is needed to determine the safety of AVs and the expected timeline of the regulatory process

- what the biggest issues in the insurance industry related to AVs are
- whether there is an emerging conflict between AV proprietary data and data that the insurance industry need.

In response to the question about whether autonomous vehicles were a game changer, the panelists concluded—for different reasons—that AV technology may not be transformative for liability and insurance regimes in the near term. For example, panelists discussed the presence of 260 million non-AVs on the road today, and that they will not go away overnight. One panelist explained that the Insurance Institute for Highway Safety predicts a three-decade adoption period between the introduction of a safety feature (e.g., antilock brakes) and its adoption. He pointed out that there are going to be "refusniks" who will not adopt AVs, regardless of their safety benefits. The panelists discussed how AVs are currently exclusively corporate and that there is no individual ownership of AVs. There was general agreement among the panelists that, until AVs are widespread among individuals, the public will not see changes in insurance industry standards. One panelist predicted that the insurance industry will maintain the status quo until there is a full shift over to AVs—probably in 20–30 years. Such a shift could mean a change from operator to manufacturer liability, but panelists agreed that more information was needed to make such a prediction.

Another panelist responded to the game changer question by stating that we are currently in a driver-assisted world, and we will be here for a long time—that is, until compatibility issues are dealt with and infrastructure is developed. She stated that the liability regime right now is flexible enough to hold manufacturers liable for failures of their product. The panelists agreed that, when it comes to deciding who to sue in an AV crash, it will be much more difficult to figure out who is liable. The panel considered the proposition that there may not be lawsuits about AV technology unless there are catastrophic losses. One of the factors to be considered is the reality of the contingency-based system in a resource-heavy investigation. The panel debated whether juries would handle AV technologies as they have handled other new technologies in the past (e.g., air bags, tire design). A panelist concluded that the liability fears on the part of AV manufacturers are overblown and that there would be no fundamental shift concerning legal liability.

Another panelist concurred that AVs would not be a game changer in the liability world because the tort system could handle these changes. However, the panelist pointed out other areas that are more complicated to consider, such as how we pay for our transportation, whether there should be a vehicle-miles traveled (VMT) tax, and whether there will be unintended consequences associated with AV technologies. She explained that we do not know what the unintended consequences might be and noted that, in 2015, there were between 4,000 to 5,000 pedestrian deaths. She speculated whether the introduction of fully autonomous vehicles might increase pedestrian deaths. She drew a parallel with 1934, the high-water mark of pedestrian deaths, when pedestrians had not yet learned to function safely in the new world of motor vehicle traffic.

A participant in the roundtable discussion commented that pedestrians will need to be held accountable in the era of AV technology because the pedestrian issue will need to be addressed to regulate AV technology responsibly. Another participant commented that Honda developed a prototype to send a message to a car through a cell phone. If a person has a disability or simply wants to cross the street, he or she can send a message to approaching cars. According to the workshop participant, the technology is there for AVs to use. The Honda cell-phone application was created to protect road construction workers; however, this type of technology could also be used for cyclists, pet walkers, and other pedestrians.

The moderator suggested that, as a society, humans accept that they are not perfect—that they make mistakes. She also noted that, however, we seem to have a problem accepting that machines also make mistakes. She asked the panelists whether they thought that the manufacturers and creators of AVs are going to be at an unfair disadvantage in the courtroom, stemming from an anti-machine bias. A panelist responded that one of the things the ride-sharing industry has learned is that consumers are very demanding. She explained that there is an apparent cultural shift toward expecting perfection in terms of efficiency, reliability, and safety. She predicted that consumers will be extremely demanding when it comes to AV technology. Another panelist commented that, currently, juries tend to understand people making mistakes but not machines making mistakes. She concluded that, perhaps, there might be an anti-machine bias concerning AV technology. Another panelist responded that, if the societal expectation is that AV technology is going to save lives, then society is going to hold the manufacturer to that standard. If purveyors of AVs make statements about safety, then people will expect AVs to be safe.

The panel discussed what is "safe enough" for AV deployment from their perspective. Panelists were asked to respond to a quote by a Toyota Research Institute executive: "[t]o achieve full autonomy, we need trillion-mile reliability." One panelist stated that if an AV is safer than the vehicle a consumer currently drives, then that value proposition will lead to adoption. Another panelist commented that the onus should be on the manufacturer; if the AV is offered to the public, the manufacturer is vouching for its safety. A panelist commented that, for consumers, it is not about the number of miles. It is about the transparency of the developers because consumers want to know how things work. A panelist noted that young people today are not getting driver's licenses, which signifies a major generational shift. She explained that there are shifts occurring in the way we perceive safety as well. She stated that paradigmatic shifts are occurring rapidly, and people are more willing to change their behavior. Some panelists agreed that more people will adopt AV technology than we expect, and more quickly. Another panelist countered that the difference with the AV is that people will expect complete safety and will not think they are assuming any risk.

The discussion turned to the regulatory process for AVs and the estimated timeline for regulatory changes. The panel identified factors that could impact the regulatory timeline, including interaction between Congress and regulatory agencies, the proliferation of current proposed legislation about AVs on Capitol Hill, and the development of standards by the National Highway Traffic Safety Administration (NHTSA). A panelist was deeply concerned that the House Committee on Transportation, and not NHTSA, was writing rules for AV technology. She stated that rules should be written by career regulators who are designing safety regulations in an apolitical environment. One of the workshop participants commented that NHTSA has not yet issued standards, probably because the agency wants to "aim first, then shoot." He observed that NHTSA was doing a variety of research projects—including human factors research—and trying, within a limited budget, to study these issues and work with auto manufacturers.

Panelists were asked to identify the biggest issues in the insurance industry related to AVs. The issues identified ranged from the inability of the insurance industry to measure aspects of AV technology, the need for regulatory changes, and the potential need to insure against AV hacking by malicious actors. For example, a panelist explained that, if a foreign power hacked all AVs in the United States and had them all turn left at the same time, it might require catastrophic event insurance, such as flood insurance. Another panelist stated that uniform national standards were going to be essential and that we may need regulatory changes that would allow these standards to develop quickly and efficiently. He pointed to regulation of transportation network companies (TNCs) as an example. He explained that the insurance industry and TNCs (e.g., Uber, Lyft) worked together to create a regulatory regime that works to cover this new industry. Another panelist commented that insurance was the biggest obstacle to TNC industry success. However, if AV technology performs as promised, panelists agreed that car insurance will become less expensive. A participant in the workshop commented that the panelists had missed a major issue: consumers and consumer protection. She noted that consumer ignorance is a major issue and provided the example of a retiree who assumes that his AV will take care of him, not understanding that an emerging technology may not always do so.

A workshop participant raised a point to the panel, stating that the number-one public policy issue for the insurance industry is the relevance and importance of data access. He explained that data help build out the framework of regulation and confirm that the technology works as advertised. However, when the insurance industry raises the issue of access to AV data, he said that the response from manufacturers is "that's proprietary." Yet, he said, data are the key to everything, and it is impossible to tease apart safety data from other data that are important to companies on a proprietary level, such as data usage. He concluded that all stakeholders are thinking about data—and everyone is guarding them. Panelists and workshop participants discussed the importance of data and data privacy for AVs. A participant commented that, because of data, we know that airbags are saving lives and reducing injuries. He explained that the insurance industry is worried about the emerging conflict between proprietary data on one hand and the data that the public—or insurance company—needs on the other. He noted that, currently, this issue is most visible in the TNC industry, but it will be a big problem when AVs are on the road.

Another participant pointed out that insurers will need data to pay claims, for example, to answer a question about whether the AV stopped improperly. He explained that, for the insurance world, data are used to resolve claims. He noted that the data may be proprietary, but they are still needed to pay off claims. In the end, he said, it is a consumer issue because claims will need to be paid. Another workshop participant agreed, pointing out that insurance contracts for her company include a duty to cooperate with the insurer in resolving a claim. She said that would not change just because AVs were involved. A panelist pointed out that it is in everyone's best interest to get claims processed efficiently. He added that it was incumbent on the insurance industry to determine how to reduce friction in the claims system. Another panelist added that it is important for regulators to preserve the ability to aggregate data, pointing out that, in the past, State Farm was able to see trends in data concerning the dangers of Firestone tires before NHTSA was aware of any safety problems. A participant closed the discussion by observing that the issue of data is not new. She explained that there have been black boxes on airplanes for many years, and their data have been used in accident reconstruction. The challenge will be how to open the door to important data from an automotive standpoint.

The panel discussion concluded with panelists being asked to briefly summarize the critical challenges posed by AVs in the next five years. Two panelists concurred that knowing about these challenges and understanding AV technology were critical. Another panelist suggested that putting a mechanism in place to deal with unexpected consequences was critical, as was assuring that there is flexibility in the regulatory framework to accommodate unexpected results. The opportunity to educate consumers was proposed as a critical challenge by another panelist. The panelists agreed that safety would be the cornerstone for any of these important challenges.

3. Lunch Presentation: The Conundrum of Autonomous Vehicle Safety

Nidhi Kalra, Director, San Francisco Bay Area Office, RAND Corporation

The lunch presentation began with three questions. First, how safe should autonomous vehicles (AVs) be? Second, how can we know how safe they are? Third, how do we design good AV policy, particularly given the answers to the first and second questions? To answer the first question, the speaker asked workshop participants to consider a statement released by University of Michigan researchers: "For consumers to accept driverless vehicles, the researchers say tests will need to prove with 80 percent confidence that they're *90 percent safer than human drivers*" (Carney and Moore, 2017). The presentation addressed how consumers may evaluate whether AVs are safer than human drivers. Participants were introduced to a spectrum of safety performance, from perfection (no crashes) to not very good performance, which was left undefined. Somewhere in the middle were average human drivers. There were three conceptually important points along this spectrum that indicated opportunities for introducing AVs: near perfect, better than the average human, and not as good as the average human driver. Each of these points presented unexpected advantages and drawbacks. For example, if society waited for nearly perfect AVs before introducing them into the market, the rate of lives lost would eventually drop to nearly zero. But initially, few lives would be saved because it would be many years before these vehicles would be widely, then fully, adopted. The speaker explained that number of lives saved with and without this technology should not to be measured solely at the point of introduction but a time in the future when adoption is widespread.

The speaker proposed that, alternatively, society might wait until AVs are better than the average human driver. Presumably, this would occur before the technology is perfect. Initially, the number of lives saved might not be significant because adoption would be slow—possibly slower than with the near-perfect technologies because of their lower performance. However, it is conceivable that AVs might be improved faster when widely deployed than they would if they remained test vehicles because developers and regulators could have much more data from which to learn. Even if AVs did not reach near-perfection any sooner, their earlier deployment would mean more lives saved in aggregate.

Participants were asked to consider what might happen if society adopted the technology even earlier, before it is as good as the average human driver. The technology could be deployed sooner, but at the expense of more crashes, as least initially. Counterintuitively, the speaker explained, more lives might be saved with this "not-quite-there" standard if developers can use early deployment to rapidly improve the AVs. AVs might become at least as good as the average human faster than they might otherwise and, thus, save more lives overall. On the other hand, she conceded, public backlash from the inevitable crash from a not-quite-there technology might be so great as to put a stop to the industry.

There is a utilitarian argument that AVs should be allowed as soon as they are safer than the average human driver, so society can start reducing the enormous loss of life from human drivers. Although people will tolerate mistakes from other people, they are far less forgiving when machines make mistakes. Some consumers will feel that until AVs are nearly perfect, they should not be on our roads. The speaker concluded that there is no correct answer; it is a difference of values and culture. But it means the answer to "how safe" AVs are remains uncertain.

Workshop participants were invited to consider the second question: "How can we know how safe they are?" The RAND report, *Driving to Safety: How Many Miles of Driving Would It Take to Demonstrate Autonomous Vehicle Reliability?* (Kalra and Paddock, 2016), showed that society cannot use test driving as a means of proving that AVs are safe because such proof requires hundreds of millions—if not billions—of miles of driving. There is a push to develop modeling and simulation methods, test courses, and other means, but these are not yet developed and validated. The result is that the answer to the second question also remains uncertain.

The third question addressed the design of good AV policy. The presentation examined how Federal Motor Vehicle Safety Standards (FMVSS) pose a barrier to deploying some types of AVs. FMVSS specify the design, construction, performance, and durability requirements for motor vehicles. For example, the FMVSS include specifications for the size of a rearview mirror. These specifications may obstruct innovative AV designs—including vehicles without a driver. However, there is an existing exemption process that would provide a path around FMVSS to facilitate innovation. Developers may apply to the National Highway Traffic Safety Administration (NHTSA) for exemptions on several bases, including that they are developing innovative safety features. Most AV exemptions would be based on this exemption. In granting an exemption, NHTSA limits the exposure to risk in two ways. First, the developers need to demonstrate that their design is just as safe as vehicles that do conform to the standards. Second, the vehicle developers are limited to selling 2,500 vehicles per exemption per year.

For comparison, in 2016, more than 17.5 million new cars and trucks were sold in the United States, and the most popular car model had sales of nearly 400,000 vehicles (Zhang, 2017; "2016 Auto Sales . . . ," 2017). Comparatively, an exemption for 2,500 vehicles (in the case of AV exemptions) is a fairly small number. On June 20, 2017, 14 AV bills were introduced in the U.S. House of Representatives. One of these was the Practical Autonomous Vehicle Exemption Act, which proposed to raise the exemptions limit from 2,500 vehicles per exemption per year to 100,000.[1] The speaker explained that unfortunately, neither the letter nor the spirit of the exemption process works very well when it is applied to AVs. The proposed bill provides that

---

[1] The Practical Autonomous Vehicle Exemption Act became part of H.R. 3388, which was passed by the U.S. House of Representatives on September 6, 2017.

automakers must demonstrate that the nonconforming vehicle can be as safely driven as a conforming vehicle. Applied literally for AVs, if a developer wants to replace a rearview mirror with a camera, the developer must show that the camera provides the same field of view to the perception software that the mirror provided to the human driver. But this ignores the real question of how well the vehicle perceives its environment. There is no analogue in FMVSS to the design and performance of perception software, since perception is a human function governed by state regulations administered at the Department of Motor Vehicles. The speaker concluded that, in brief, FMVSS do not address the *autonomous* aspect of an AV.

The speaker explained that whether the subject is AVs or pharmaceuticals or geoengineering, there is usually a trade-off between risk and uncertainty. The more precisely we want to know how safe a disruptive, hard-to-predict technology is, the more risk we need to accept in deploying it to find out. RAND's *Driving to Safety* report examines "how many miles of driving do you need to prove AV safety" (Kalra and Paddock, 2016). It concludes that there is a trade-off: The smaller the difference, the more miles needed; the rarer the event (e.g., fatalities), the more miles needed. This means that society would have to accept a lot of risk to detect even modest improvements in AV fatality rates over human-driver fatality rates. For example, a 20-percent improvement would require approximately 5 billion miles of driving.

The presentation explored whether we should use the relationship between risk and uncertainty to limit our risk while increasing our knowledge gradually. Participants were asked to imagine a series of performance-based gates for the introduction of AV technology. At each gate, more vehicles would be allowed to be deployed, provided they demonstrably met some safety standard. This safety standard would become more and more comprehensive at each successive gate. The number of vehicles and the safety standard would be based on the trade-off between risk and uncertainty. Participants were then invited to imagine that 5,000 AVs were initially allowed per FMVSS exemption. At 12,000 miles, the AVs would collectively have driven 60 million miles in a single year. That would be sufficient mileage to determine in the first year if the AV fleet is at least 10-percent safer than human drivers in terms of crashes or at least 15-percent safer in terms of injuries. We would not know whether the AVs are better or worse in terms of fatalities unless the performance difference was very large—at least 80 percent better or, unfortunately, 80 percent worse. The speaker proposed that this might be the uncertainty society would accept to limit risk.

Finally, workshop participants were invited to suppose that 20,000 AVs were deployed per FMVSS exemption. When the AVs had driven 12,000 miles, that would provide a collective amount of 240,000 miles for gaining information. It would allow experts to detect essentially any difference in crashes and injuries, but importantly, it would allow detection of differences in fatality rates of 55 percent or more. The speaker proposed that society might adopt this level of risk as a first performance gate so that we can learn more, or it might be the basis of the second performance gate once the first one has been cleared.

The speaker concluded that, when facing great uncertainty, it is almost impossible to get regulations right the first time. Policies that manage uncertainty could enable innovation, while balancing the trade-off between risk and information.

4. Afternoon Breakout Sessions: Liability and Insurance Rethink

In the afternoon, the participants were divided into two smaller working groups designed to facilitate deeper conversation while tackling several different questions and problems related to liability and insurance for autonomous vehicles (AVs). Both groups were asked to identify important research questions in these areas. Significantly, both groups identified multiple areas for research. The deliberations of both groups have been combined in the discussion below.

Insurance Industry Challenges from Autonomous Vehicles

There was a robust discussion surrounding how insurance companies should determine the cost of AV insurance plans and decide how to pay out following AV accidents. Currently, an insurance company's analysis of a car accident between two standard vehicles does not involve the manufacturer of the cars. If, in the future, the accident is between two AVs, the current approach will not work. It is infeasible to subrogate each accident, which would be extremely inefficient. Participants debated whether they would adjudicate AV claims against Volvo (which claims 100 percent responsibility for its AVs) differently from Tesla (which asserts 0 percent responsibility). The group agreed that having these numbers determined in advance would make the process much more efficient. One potential solution the group proposed was to have annual contracts between insurance companies and original equipment manufacturers (OEMs) that state, "if there is an accident, and we are responsible, we will pay X amount." Insurance companies could then decide what to charge customers based on the AV they decide to "drive." Insurance companies would determine fault confidentially, almost similar to confidential arbitration. The group compared this approach with no-fault liability schemes.

In contrast, several members of the group suggested that the technology might make traditional accident-by-accident adjudication more efficient than it might appear. For example, "black boxes" of data on AVs, similar to an event data recorder, when paired with emerging technology, might allow a neutral third-party algorithm to determine fault. Another solution that was proposed was to stipulate that, if an AV is out of its lane on the road, the owner would be responsible for an accident (a similar approach to how insurance companies currently handle rear-end collisions). Insurance pricing was also discussed at length. The group discussed how insurer metrics have always been driver-focused and the potential metrics insurers could use to price insurance for AVs. Several members of the group explained why insurance companies need access to manufacturer data to make underwriting decisions. The group debated how to create insurance pricing for semi-autonomous cars where the driver can elect to use self-driving capabilities. Another interesting issue was how to provide coverage for malicious intervention with an AV by a hacker. Currently, there is no tort coverage in insurance policies (e.g., for pain and suffering) if the victim cannot identify the hacker.

Data Sharing and Privacy Issues

The group discussed the tensions between sharing data for safety and insurance reasons and protecting consumer privacy. One of the most important privacy issues is de-identifying location and geographic data. The industry needs to reevaluate what types of data are needed from an AV, as opposed to a traditional automobile. The insurance industry should participate in data sharing, which is incumbent on all stakeholders. Data sharing between manufacturers and insurance companies is necessary for insurance pricing; data sharing between manufacturers and courts is necessary for litigation; and data sharing between manufacturers and law enforcement is necessary for security and crime prevention and prosecution.

Artificial Intelligence and Evidence in Litigation

The group agreed that it was important to understand the state of the art of artificial intelligence (AI). A key concern was the lack of traceability of AI. The group discussed the need for manufacturer documentation of AI used in an AV, as well as the potential for strict liability because of low traceability. One of the issues in litigation will be explaining the data gathered from an AV. Expert witnesses will be needed to explain the state of the art of AI technology. A related question will be: "Who is responsible for educating the consumer about AI?"

Job Loss, Innovation, and the Taxi and Trucker Question

The group agreed that the conventional wisdom is that driving jobs will disappear with the full deployment of AVs, but they questioned if this was accurate. The group debated whether AVs might actually increase opportunity for more-experienced drivers, such as taxi drivers. The opportunity for experienced drivers was discussed in the context of the average person not being able to drive certain cars, including AVs. The group discussed the possibility of retraining the taxi- and truck-driver workforces as technology advances. A participant in the group described how Allstate recently implemented pilot programs in Texas and California—which were then rolled out nationwide—that fundamentally altered the Allstate claims process. In the past, Allstate customers had to drive their recently crashed vehicles to an Allstate claims center for damage assessment. Now, customers can take a photograph of the damage with their smartphone while the car is in their own driveway and send the photo and claim to Allstate through a smartphone application. An automobile-claims process that used to take five to six hours has been reduced to minutes. The group discussed how this improvement has not meant that Allstate's claims adjusters have lost their jobs. Instead, claims adjusters were retrained to use the new technology. This may be the future for some truck drivers—for example, drivers who can handle *truck platooning*—autonomous trucks that follow one another very closely in a convoy on a major highway. Group participants discussed how trucking companies are currently exploring platooning as a mechanism for saving fuel costs, rather than labor costs, and there may be aspects of platooning that will require a human in the truck.

California's Proposition 103

The group discussed the probability that the insurance industry will not be able to insure AVs in California because of Proposition 103, which requires that insurers focus on three factors when determining the cost of coverage: (1) driving record, (2) number of miles the driver has driven, and (3) length of time the driver has been driving. Clearly, these criteria are inapplicable in an AV world. An insurance industry group has been working on educating lawmakers, but it is a politically sensitive subject. Consumer watchdog groups are generally 100 percent opposed to AVs, according to a group participant. The group explored potential solutions, including whether Proposition 103 definitions could be tweaked and whether Level 4 (high automation) and Level 5 (full automation) vehicles might fall outside Proposition 103 entirely.

Importance of Sound Infrastructure

The group agreed on the essential role of infrastructure. A group participant suggested that stakeholders create an infrastructure wish list. Specifically, stakeholders should identify the actions that should be taken by local, state, and federal governments to optimize AV technology. The group identified grading roads, optimizing road conditions, improving signage and striping, and creating a standard classification system as important infrastructure issues. One group member noted that, perhaps, AV stakeholders do not want anything new; they just need the government to fix broken infrastructure and focus on good maintenance. In a related discussion, members of the group pointed out that too much of the burden concerning AVs is being placed on manufacturers. They suggested that, in addition to focusing on infrastructure, there should also be attention focused on prevention of jaywalkers, laws regulating driving speeds, and so forth.

Regulation and Unexpected Consequences

The group debated the government's role in managing the competing types of technologies on the road. A key issue the group identified concerned whether ridesharing would increase the number of cars on the road. The group concurred that there could be an increase in miles driven, given the ease of use of ridesharing. One group participant described a report prepared by the U.S. Department of Energy that calculated that ridesharing could result in a decrease of miles driven by 60 percent or an increase by 200 percent. The group discussed how we, as a society, can respond quickly to unexpected consequences from AV technology that we were unable to plan for. Perhaps policies and practices that have been politically unpalatable in the past might be possible now, since they are coming from the private sector, rather than the government. For example, Uber's surge pricing may be just a congestion tax by another name. The group agreed that there may be other policies that AVs make newly possible.

5. Key Issues, Research Questions, and Next Steps

In the final session, each of the working groups reported out their key issues based on the group discussions and provided suggestions for further research. This sparked a roundtable discussion of both the issues and of topics that require further research.

The groups highlighted the following issues as the most critical:

- **Insurers and manufacturers need to identify the "state of the art" of autonomous vehicle (AV) technologies going forward.** For example, in the litigation context, it will be critical to identify what was known by the AV manufacturer at a given time and the state of the art of AV technologies at the same given time. Litigation probably will occur three to four years after an AV crash, so information about the state of the art of the technology at the time of the event, rather than currently, will be necessary.

- **AV manufacturers, developers, and insurers need to fully understand artificial intelligence and explain to consumers how AI makes decisions.** It is unclear whether consumers will demand this information prior to adopting AV technologies. Perhaps younger people may simply adopt new technologies without transparency concerning how the technology works.

- **Insurers need AV data to create pricing policies for automobile insurance and to assess underwriting risk.** It is unclear how insurers can obtain the needed data from AVs. Manufacturers are guarding most of AV data as "proprietary." Consumers may not have the right to disclose the data, although in the future, data might be required as a condition of insurance. Perhaps, in the future, there will be a compromise between insurers and manufacturers. It is unclear what role, if any, the government will play in the AV data-sharing process. Privacy will be a key issue, and whether and how data can be successfully de-identified will be critical. Accident data and attribute data are both relevant to the insurance industry.

- **Security of AVs, particularly in terms of "hacking" by a malicious actor, is a very important issue for the insurance industry.** There are dangers posed by massive hacks of AVs, as well as private hacks, which are random and more isolated. Possibly, insurers can identify opportunities for new insurance products. Currently, auto insurance policies may not cover either massive or individual hacking of AVs.

- **Original equipment manufacturers (OEMs) have powerful brands and are going to hesitate to admit fault in the event of an AV crash.** This may create complications in resolving insurance-related conflicts between AVs manufactured by different OEMs.

- **The advent of new technologies might allow us to rethink how our traditional approach to regulation may not create good policy.** For example, use-based fees might become acceptable.

- **Predictions of catastrophic job loss resulting from AVs may be overstated in the short term.** It is probable that highly trained drivers will be needed to monitor AVs initially. For example, truck platooning may result in fuel savings that could help keep driver's wages high.

- **It is important to "think beyond the car" in creating a safe environment for the introduction of AVs.** Stakeholders should obtain partners to achieve better road infrastructure, markings, and signage, as well as regulations to prevent jaywalkers and other pedestrian hazards.
- **It may be difficult to insure AVs in California under the current regime unless Proposition 103 is changed.** Proposition 103 requires that insurers focus on three factors when determining the cost of coverage: (1) driving record, (2) number of miles the driver has driven, and (3) length of time the driver has been driving. These criteria will be inapplicable in an AV world.

The two groups identified many issues for research. These included the following:

- **Investigate and better understand consumer attitudes about trusting and adopting a new technology.** This research should identify and explore consumers' expectations of safety and identify any generational discrepancies. The research might include an evaluation of how frequently optional safety features are purchased and used, how frequently technology is blamed—incorrectly—for automobile accidents, and whether language barriers inhibit the explanation of new technology to consumers.
- **Create a research design to evaluate consumers' use of AV technology as soon as it is available to consumers.** The research design would address the extent to which consumers follow manufacturer guidelines and instructions regarding safety and use of an AV, how soon after purchasing an AV are consumers involved in an accident, and the types of training manufacturers provide to consumers at the time of an AV purchase. The research could also address the degree of understanding of AV technology that consumers need by tracking and evaluating AV product liability litigation.
- **Conduct research to explore other industries that have experience with AI (e.g., aviation, nuclear engineering) to identify how the AV industry might learn from them.** As part of this research, evaluate frameworks for redundancy used in other industries that would be applicable to AVs.
- **Evaluate whether driving culture varies by region and geography and whether AV algorithms will need to drive differently in different regions.** Another aspect of this research could assess how the gradual deployment of AVs will impact individual access to transportation in different regions of the United States.
- **Investigate the AV data that are needed by consumers, manufacturers, regulators, and insurers.** Assess the best practices to collect and verify AV data. Explore how to measure the safety benefits of AVs and how these findings may be used to improve road infrastructure. Identify the changes to FMVSS and other safety standards that should apply to AVs. Attempt to quantify the percentage of the fleet that will need to be autonomous to produce the societywide benefits promised by the new technology. Track AV data to demonstrate the lag time between the purchase of an AV and the first accident.
- **Conduct research to identify and quantify the impact of AVs on insurance coverage.** The research might explore how the insurance industry can deal with the proliferation of actors involved in accident liability and address whether the insurance industry would be best served by handling liability issues proactively through consumer contracts or reactively through the traditional tort system. This analysis might consider the difference between everyday driving and high-exposure situations.

- **Evaluate cyber safety and privacy issues posed by AVs.** Evaluate the potential trade-off between the benefits of shared AV data (e.g., between manufacturers and insurers) and consumers' reasonable expectation of privacy. The research could identify the risks posed by AI and attempt to determine who might be held liable for accidents or damages resulting from faulty AI.

The workshop concluded with a final roundtable discussion prompted by the research questions identified by both groups. Some of the key ideas for next steps that emerged in this discussion were suggestions that the insurance industry needs to be proactive, not reactive. The group discussed the possibility of developing new insurance coverage models and the need for a working group to further these efforts. A California resident thought that an AV working group should consider how to fix Proposition 103 so that the criteria could apply to AVs. In addition, the group discussed developing a time frame for regulators and Congress about the development of AV technology. A working group might identify and address the types of international cooperation that would be possible. Generally, the group thought it was important to find a way to modernize regulations to better deal with such industry-altering technologies as autonomous vehicles. The group debated how to deal with the public-relations issues associated with the initial deployment of AVs. The group concurred that consumer acceptance and education would be essential to successful widespread deployment of AV technology. The participants agreed that they would like to find a way to continue the conversations that emerged in the day's workshop.

Institute for Civil Justice

*Rethinking Insurance and Liability in the
Transformative Age of Autonomous Vehicles*

O'Melveny & Myers LLP
Two Embarcadero Center | 28th Floor
San Francisco, CA

July 14, 2017

<u>Agenda</u>

8:30 a.m.   *Breakfast and Networking*

9:00 a.m.   Welcome and Introductions

9:15 a.m.   Setting the Stage: Why Are We Here?
*James Anderson, Director, RAND Institute for Civil Justice
Nidhi Kalra, Director, San Francisco Bay Area office,
RAND Corporation*

10:00 a.m.   *Break*

10:15 a.m.   Panel Discussion: Some Key Uncertainties that Will Shape
Future Liability for Autonomous Vehicles

Moderators:
*Karlyn D. Stanley, Adjunct Senior Researcher; Professor,
        Pardee RAND Graduate School, RAND Corporation
Sabrina Strong, Partner, O'Melveny & Myers LLP*

Panelists:
*Rena Davis, Public Policy Manager, Lyft
Ellen Partridge, Senior Law Fellow, Environmental Law & Policy Center
Christine D. Spagnoli, Partner, Greene Broillet & Wheeler, LLC
James Whittle, Associate General Counsel & Chief Claims Counsel,
        American Insurance Association*

12:00 p.m.	*Lunch and Presentation* The Conundrum of Autonomous Vehicle Safety *Nidhi Kalra*
1:15 p.m.	Breakout Groups: Liability and Insurance Rethink *Group leaders:* *James Anderson, RAND Corporation* *Richard Goetz, O'Melveny & Myers LLP* *Melody Drummond Hansen, O'Melveny & Myers LLP* *Karlyn D. Stanley, RAND Corporation*
3:00 p.m.	*Break and Networking*
3:15 p.m.	Research Questions and Next Steps *James Anderson*
3:45 p.m.	*Closing Remarks* Dan C. Dunmoyer, Chair, RAND Institute for Civil Justice Board of Overseers
4:00 p.m.	Reception O'Melveny & Myers LLP

Institute for Civil Justice

Rethinking Insurance and Liability in the
Transformative Age of Autonomous Vehicles

Attendee List
July 14, 2017

James Anderson
Director, Institute for Civil Justice
RAND Corporation

Sue Bai
Principal Engineer,
Automobile Technology Research
Honda R&D Americas, Inc.

Catherine Cruz
Development Associate
RAND Corporation

Rena Davis
Public Policy Manager
Lyft

Melody Drummond Hansen
Partner
O'Melveny & Myers LLP

Dan Dunmoyer
Chair, RAND Institute for Civil Justice
Board of Overseers;
Head of Government and
Industry Affairs, USA
Farmers Group, Inc.

Armand Feliciano
Vice President
Association of California Insurance
Companies - Property Casualty Insurers
Association of America

Ryan Gammelgard
Counsel
State Farm Mutual
Automobile Insurance Company

Richard Goetz
Partner
O'Melveny & Myers LLP

Angie Hankins
Senior Director of Strategy,
Senior Investment & IP Counsel
Samsung Electronics,
Strategy & Innovation Center

Nidhi Kalra
Director, San Francisco Bay Area office
RAND Corporation

Kenneth Korea
Senior Vice President & Head
Samsung Electronics,
United States Intellectual Property Center

Institute for Civil Justice

Sonja Larkin-Thorne
Consumer Advocate
National Association of
Insurance Commissioners

Annie Lee
Intern
O'Melveny & Myers LLP

Danielle Lenth
California Lead Legal Counsel
Allstate Insurance Company

Jamie Morikawa
Associate Director,
Institute for Civil Justice
RAND Corporation

Michael Nelson
Partner
Eversheds Sutherland LLP

Celeste Parisi
Apple

Ellen Partridge
Senior Law Fellow
Environmental Law & Policy Center

Anna Schneider
Intern
O'Melveny & Myers LLP

Christine Spagnoli
Partner
Greene Broillet & Wheeler, LLP

Karlyn Stanley
Adjunct Senior Researcher
RAND Corporation

Gary Strannigan
Associate Vice President, Public Affairs
Liberty Mutual Insurance

Sabrina Strong
Partner
O'Melveny & Myers LLP

James Whittle
Associate General Counsel &
Chief Claims Counsel
American Insurance Association

Alexander Willis
Manager of Industry Affairs
Japan Automobile
Manufacturers Association, Inc.

Stephen Wood
Acting Chief Counsel for Vehicle Safety
Standards & Harmonization
National Highway Traffic
Safety Administration

Phillip Zackler
Senior Legal and Compliance Counsel
Toyota Research Institute

References

"2016 U.S. Auto Sales Set a New Record High, Led by SUVs," *Los Angeles Times*, January 4, 2017. As of December 28, 2017:
http://www.latimes.com/business/autos/la-fi-hy-auto-sales-20170104-story.html

Anderson, James M., Nidhi Kalra, Karlyn D. Stanley, Paul Sorensen, Constantine Samaras, Oluwatobi A. Oluwatola, *Autonomous Vehicle Technology: A Guide for Policymakers*, Santa Monica, Calif.: RAND Corporation, RR-443/2-RC, 2016. As of March 20, 2018:
https://www.rand.org/pubs/research_reports/RR443-2.html

Carney, Sue, and Nicole Casal Moore, "New Way to Test Self-Driving Cars Could Cut 99.9 Percent of Validation Costs," *Michigan News*, May 23, 2017. As of December 28, 2017:
http://ns.umich.edu/new/releases/24866-new-way-to-test-self-driving-cars-could-cut-99-9-percent-of-validation-costs

Fraade-Blanar, Laura, and Nidhi Kalra, *Autonomous Vehicles and Federal Safety Standards: An Exemption to the Rule?* Santa Monica, Calif.: RAND Corporation, PE-248-RC, 2017. As of March 20, 2018:
https://www.rand.org/pubs/perspectives/PE258.html

Kalra, Nidhi, and David G. Groves, *The Enemy of Good: Estimating the Cost of Waiting for Nearly Perfect Automated Vehicles*, Santa Monica, Calif.: RAND Corporation, RR-2150-RC, 2017. As of March 20, 2018:
https://www.rand.org/pubs/research_reports/RR2150.html

Kalra, Nidhi, and Susan M. Paddock, *Driving to Safety: How Many Miles of Driving Would It Take to Demonstrate Autonomous Vehicle Reliability?* Santa Monica, Calif.: RAND Corporation, RR-1478-RC, 2016. As of December 5, 2017:
https://www.rand.org/pubs/research_reports/RR1478.html

National Conference of State Legislatures, "Autonomous Vehicles | Self-Driving Vehicles Enacted Legislation," webpage, January 2, 2018. As of January 2, 2018:
http://www.ncsl.org/research/transportation/autonomous-vehicles-self-driving-vehicles-enacted-legislation.aspx

National Highway Traffic Safety Administration, "Automated Vehicles for Safety," webpage, undated. As of December 1, 2017:
https://www.nhtsa.gov/technology-innovation/automated-vehicles-safety

NCSL—*See* National Conference of State Legislatures.

NHTSA—*See* National Highway Traffic Safety Administration.

U.S. Congress, 115th Cong., 1st Sess., Safely Ensuring Lives Future Deployment and Research in Vehicle Evolution Act (SELF DRIVE Act), Washington, D.C., H.R. 3388, September 6, 2017.

U.S. Senate, American Vision for Safer Transportation Through Advancement of Revolutionary Technologies Act (AV START Act), 115th Cong., 1st Sess., S. 1885, September 28, 2017.

Zhang, Benjamin, "The 20 Best-Selling Cars and Trucks in America," *Business Insider*, January 7, 2017. As of December 28, 2017: http://www.businessinsider.com/best-selling-cars-trucks-vehicle-america-2016-2017-1/#20-hyundai-elantra-208319-sold-during-2016-down-138-over-2015-1